A LOOK AT NATURE'S CYCLES

THE CARBON CYCLE

BY BRAY JACOBSON

Gareth Stevens
PUBLISHING

CRASHCOURSE

Please visit our website, www.garethstevens.com. For a free color catalog of all our high-quality books, call toll free 1-800-542-2595 or fax 1-877-542-2596.

Library of Congress Cataloging-in-Publication Data

Names: Jacobson, Bray, author.
Title: The carbon cycle / Bray Jacobson.
Description: New York : Gareth Stevens Publishing, [2020] | Series: A look at nature's cycles | Includes bibliographical references and index.
Identifiers: LCCN 2018043658| ISBN 9781538241103 (pbk.) | ISBN 9781538241127 (library bound) | ISBN 9781538241110 (6 pack)
Subjects: LCSH: Carbon cycle (Biogeochemistry)--Juvenile literature. | Carbon--Juvenile literature.
Classification: LCC QH344 .J33 2020 | DDC 577.144--dc23
LC record available at https://lccn.loc.gov/2018043658

First Edition

Published in 2020 by
Gareth Stevens Publishing
111 East 14th Street, Suite 349
New York, NY 10003

Designer: Sarah Liddell
Editor: Kristen Nelson

Photo credits: Cover, p. 1 (main) pongpinun traisrisilp/Shutterstock.com; cover, p. 1 (inset) Andriano/Shutterstock.com; arrow background used throughout Inka1/Shutterstock.com; p. 5 MJTH/Shutterstock.com; p. 7 Manbetta/Shutterstock.com; p. 9 snapgalleria/Shutterstock.com; p. 11 Wonderwall/Shutterstock.com; p. 13 Jon Schulte/Shutterstock.com; p. 15 MYP Studio/Shutterstock.com; p. 17 Mikhail Starodubov/Shutterstock.com; p. 19 Maksimilian/Shutterstock.com; p. 21 fboudrias/Shutterstock.com; p. 23 Ziablik/Shutterstock.com; p. 25 demamiel62/Shutterstock.com; p. 27 Marten_House/Shutterstock.com; p. 29 Rawpixel.com/Shutterstock.com; p. 30 Vecton/Shutterstock.com.

Printed in the United States of America

CPSIA compliance information: Batch #CS19GS: For further information contact Gareth Stevens, New York, New York at 1-800-542-2595.

CONTENTS

Words in the glossary appear in **bold** type the first time they are used in the text.

LIFE'S ELEMENT

One of the most important elements on Earth is carbon. Without it, life wouldn't exist! Much of the energy people use comes from carbon. Carbon is recycled through **processes** that are both naturally occurring and **influenced** by human activities.

MAKE THE GRADE

By **mass**, about 18 percent of the human body is made of carbon atoms.

ALWAYS MOVING

Carbon is always moving between living things and nonliving things. It moves from Earth into the atmosphere, or the gases that surround Earth. There are two main pathways carbon follows. These interconnect and together are called the carbon cycle.

MAKE THE GRADE

Without human activities, the natural carbon cycle is what keeps the levels of the gas carbon dioxide (CO_2) **stable** in Earth's atmosphere.

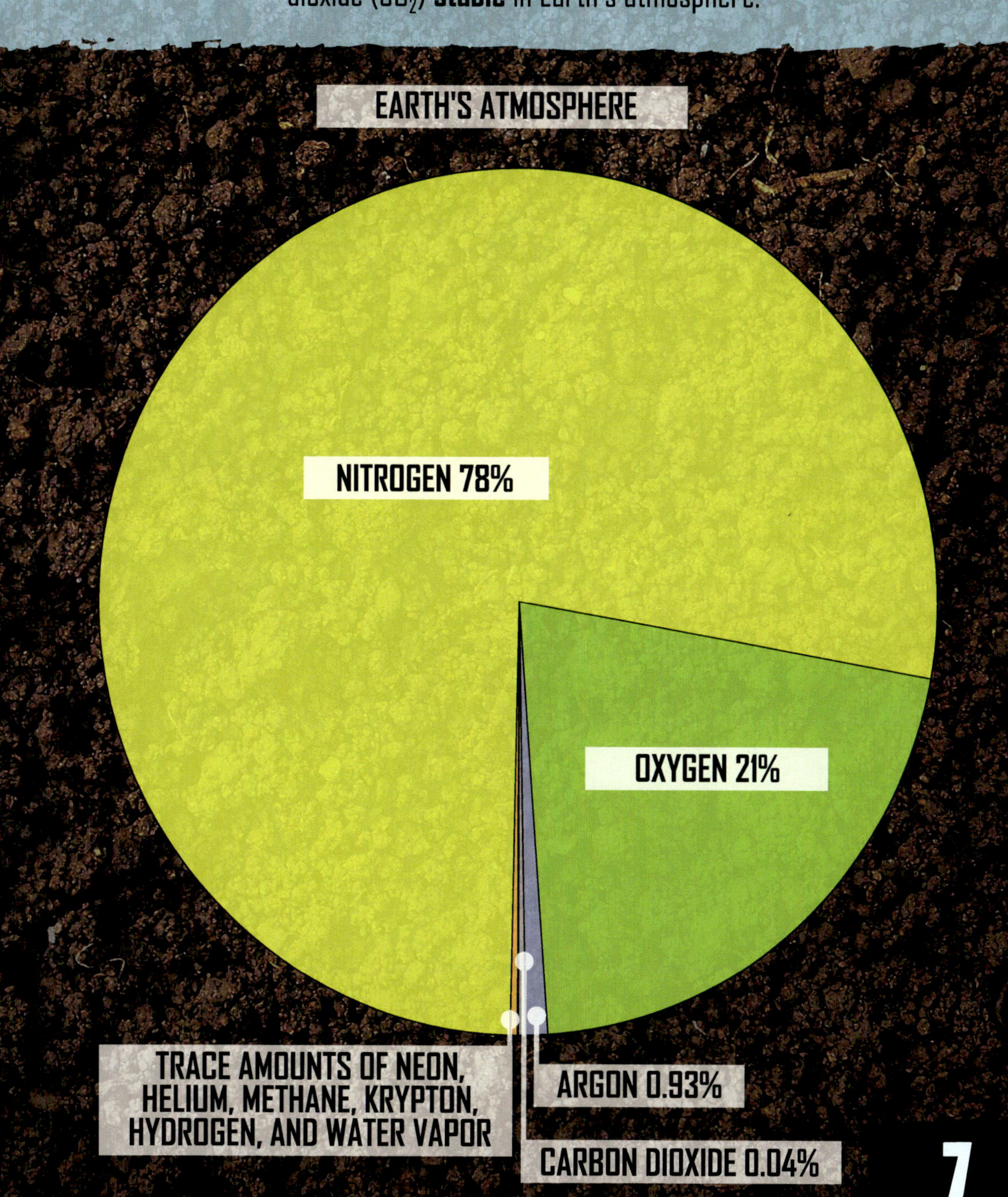

START WITH PLANTS

One part of the carbon cycle begins with plants and other **autotrophs**. They take in carbon from the atmosphere as carbon dioxide. They use it, along with water and sunlight, in a process called photosynthesis to make oxygen and their own food, called glucose.

MAKE THE GRADE

Carbon can move through the **biological** part of the carbon cycle fast!

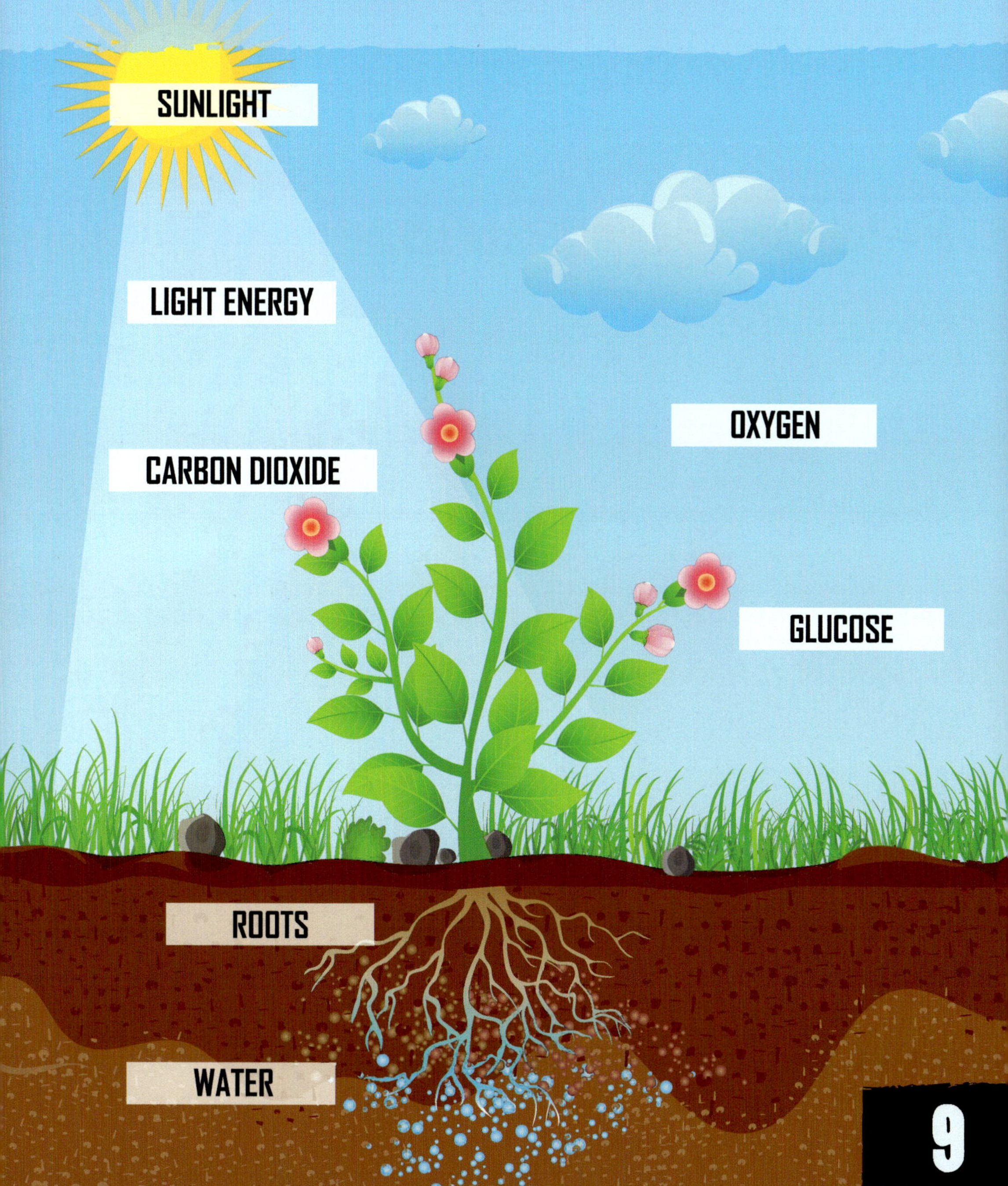

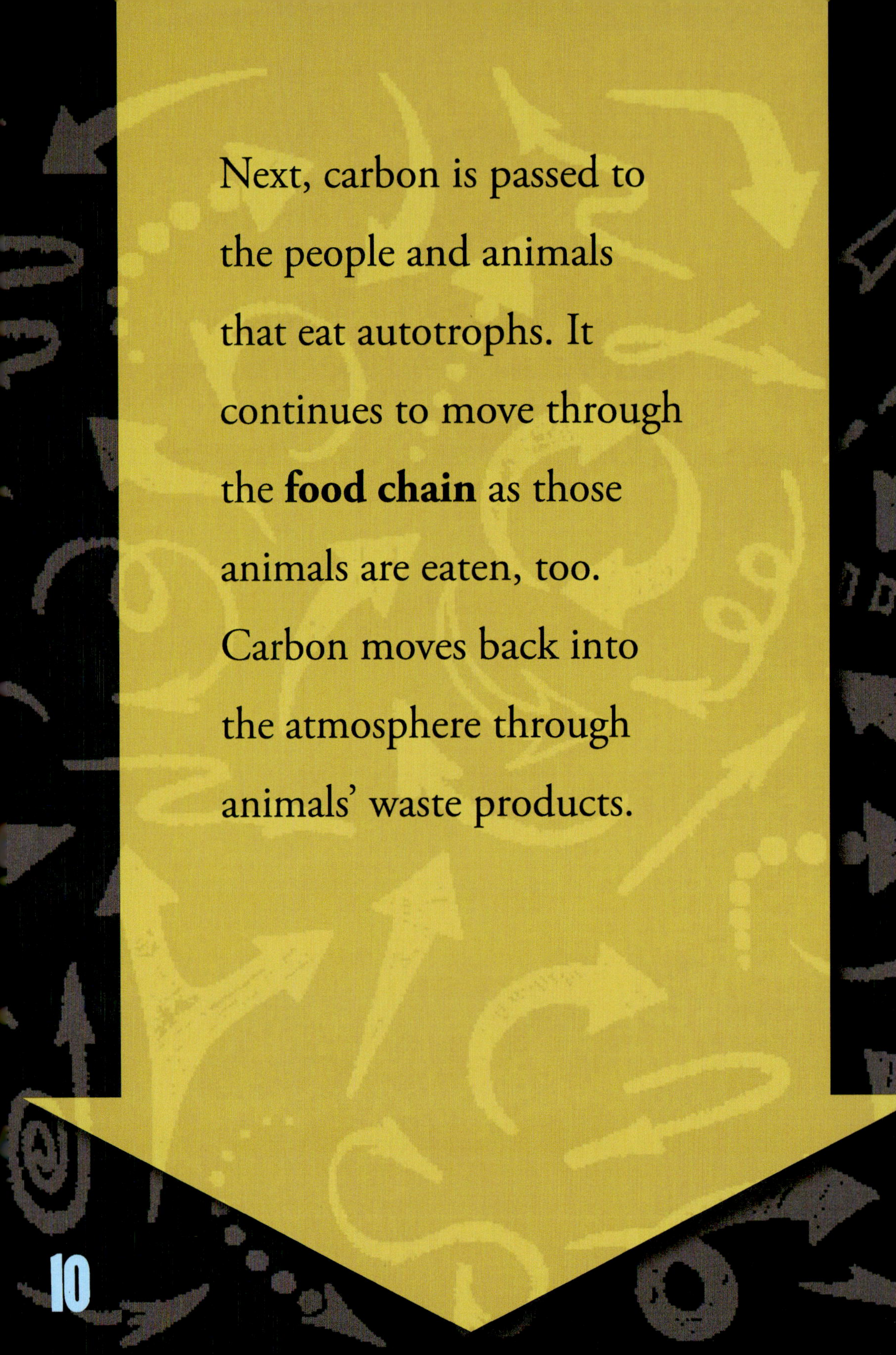

Next, carbon is passed to the people and animals that eat autotrophs. It continues to move through the **food chain** as those animals are eaten, too. Carbon moves back into the atmosphere through animals' waste products.

MAKE THE GRADE

Carbon dioxide is one of the waste products of cellular respiration. This is the process by which animal bodies and plants break down food by combining it with oxygen to **release** energy.

Once animals and plants die, they break down, or decompose. Decomposition also causes the release of carbon dioxide, putting carbon back into the atmosphere. Once this occurs, the carbon cycle can begin again.

MAKE THE GRADE

Decomposers are living things that aid in decomposition. They include fungi, bacteria, and animals like slugs. As decomposers break down plant and animal remains, carbon dioxide is released.

STORED IN EARTH

Another part of the carbon cycle is geological, which means it has to do with rocks, soil, and landforms. As living things decompose, their remains become part of the soil. This is one way carbon is stored on Earth.

MAKE THE GRADE

The geological part of the carbon cycle occurs over millions of years.

Over a long time, the remains of plants and animals are buried and **compressed**. The carbon that was in their bodies becomes stored in Earth's rock. Much of Earth's carbon is stored this way.

MAKE THE GRADE

Limestone is a kind of **sedimentary rock** that has a lot of carbon stored in it.

The remains of plants and animals buried deep within Earth become fossil fuels. These fuels—which include oil, coal, and natural gas—have a lot of carbon in them because of what they're made from! They're burned to release energy for people to use.

MAKE THE GRADE

The burning of fossil fuels is one way that carbon is released (as carbon dioxide) into the atmosphere from the geological part of the carbon cycle.

RELEASING CARBON DIOXIDE

Burning fossil fuels is a fast way to release stored carbon. A **volcanic eruption** is another quick way carbon moves from rock to the atmosphere quickly. As rock naturally wears away, carbon can be released back into the atmosphere, too.

MAKE THE GRADE

Only a small part of the carbon on Earth moves through the carbon cycle over the course of a year.

IN THE ATMOSPHERE

The carbon cycle cannot balance the rising amount of carbon in Earth's atmosphere. This is mostly because of human activities, especially the burning of fossil fuels. Carbon dioxide is a greenhouse gas, which means it traps heat from the sun in Earth's atmosphere.

MAKE THE GRADE

The ocean stores more carbon than the atmosphere. Carbon moves between the ocean and the atmosphere, too.

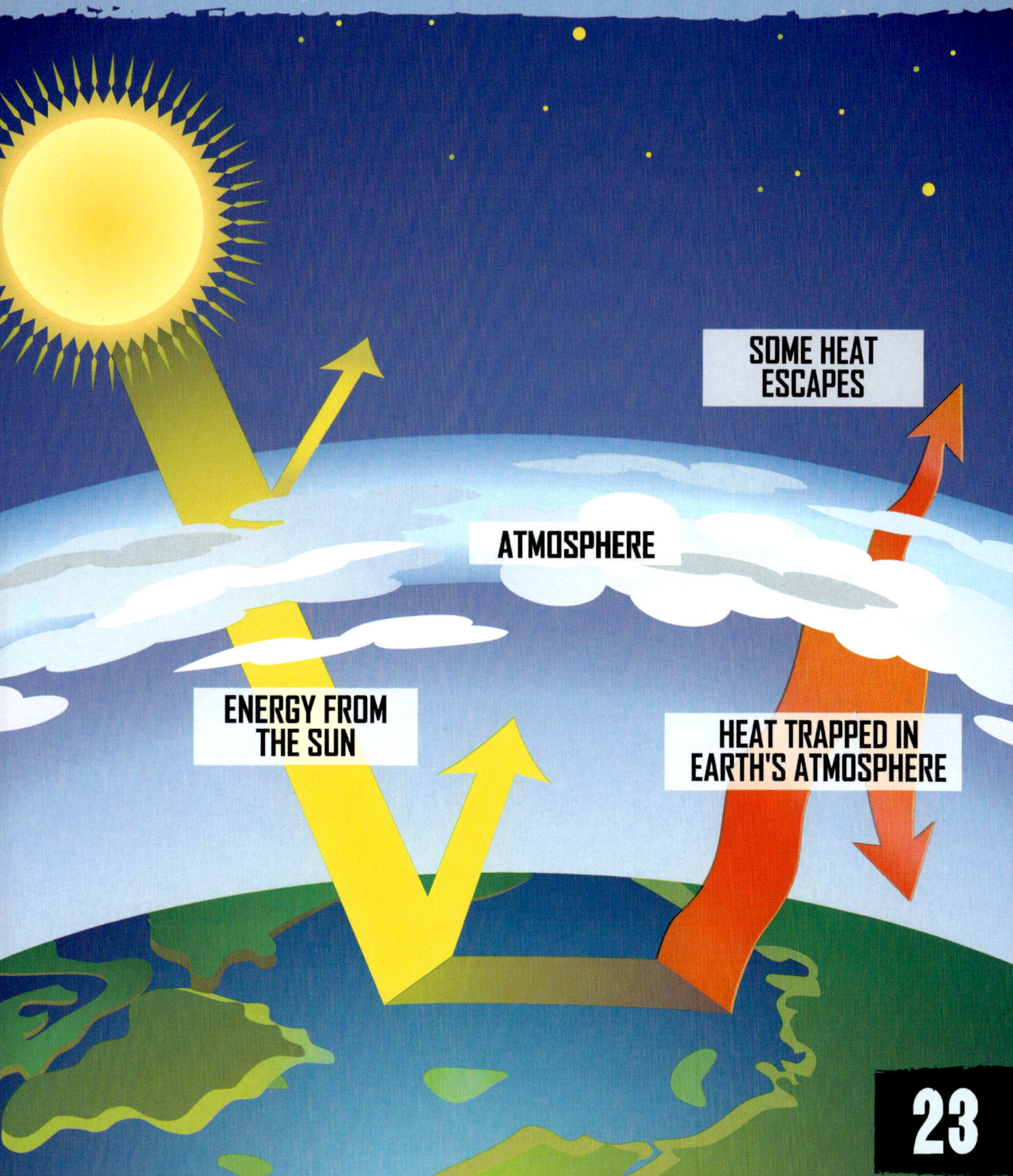

Greenhouse gases in the atmosphere are causing the average temperatures on Earth to rise. This is causing ocean levels and temperatures to rise. The more carbon dioxide is released into the atmosphere, the more the amount of carbon dioxide in the oceans will increase along with it.

MAKE THE GRADE

Earth's rising temperatures are harming ocean life.

Deforestation, or cutting down a large area of trees at once, is another human activity affecting the carbon cycle. There are fewer trees to take in the extra carbon dioxide in the atmosphere. In addition, their remains can release more carbon!

MAKE THE GRADE

Both plants and the ocean can take in some of the extra carbon dioxide in the atmosphere. But, there's too much carbon dioxide for them to balance it out.

KEEP CYCLING!

Even though human activities are causing more and more carbon dioxide to be released into the atmosphere, the carbon cycle is still at work! As long as there is life on Earth, the carbon cycle will continue to work to keep the amount of carbon balanced.

MAKE THE GRADE

To decrease the amount of carbon they use, people are asked to drive less, use less energy, and plant more trees!

THE CARBON CYCLE

CARBON DIOXIDE (CO_2)

RESPIRATION

RESPIRATION

BURNING FOSSIL FUELS

PHOTOSYNTHESIS

RESPIRATION

ANIMALS

PLANTS

DECAY

FOSSIL FUELS

GLOSSARY

autotroph: a living thing that can make its own food from nonliving matter

biological: having to do with life and living things

compress: to press or squeeze together

food chain: a way to describe the feeding relationship between animals in which each one uses the next lower member of the series for food

influence: to have an effect on

mass: the amount of matter in an object

process: a series of steps or actions taken to complete something

release: to let out or give off

sedimentary rock: the rock that forms when sand, stones, and other matter are pressed together over a long time

stable: not likely to change suddenly or greatly

volcanic eruption: the bursting forth of hot, liquid rock from within Earth

FOR MORE INFORMATION

BOOKS

Conklin, Wendy. *Earth's Cycles*. Huntington Beach, CA: Teacher Created Materials, 2016.

Loria, Laura. *The Carbon Cycle*. New York, NY: Britannica Educational Publishing, 2018.

WEBSITES

The Carbon Cycle
studyjams.scholastic.com/studyjams/jams/science/ecosystems/carbon-cycle.htm
Watch a video about the carbon cycle's role in climate change.

Greenhouse Gases and the Greenhouse Effect
kids.niehs.nih.gov/topics/natural-world/greenhouse-effect/index.htm
Find out more about the greenhouse effect and climate change.

INDEX